Physics—The Big Problem—Solved!

Physics—The Big Problem—Solved!

W.S. Oakley

VANTAGE PRESS
New York

FIRST EDITION

Published by Vantage Press, Inc.
516 West 34th Street, New York, New York 10001

Manufactured in the United States of America
ISBN: 0-533-13451-X

Library of Congress Catalog Card No.: 99-98155

0 9 8 7 6 5 4 3 2 1

Contents

Preface

While the science of physics is largely a story of success, it is evident huge issues remain unresolved. The current and very successful "Standard Model" of physics addresses many properties of photons, electrons, and atoms via Quantum Electro-Dynamics (QED) theory, but it does not include gravity. Einstein's General Relativity theory describes gravity, but does so on a cosmic scale, and the theory is therefore isolated from the elementary particle scale of QED. Consequently large and fundamental questions remain unanswered.

Why are QED and General Relativity not easily merged into a single Unified Theory? How does General Relativity apply to particles? Why do the strengths of the fundamental forces, the strong force, electromagnetism, and gravity differ so widely—by a factor of about 10^{42}? How do the properties of mass and charge arise at the particle scale? What is the nature of elementary particles? Why do the masses of electrons and protons differ by about 1,836? It has been over three hundred years since Newton put physics on a mathematical basis and much has been accomplished since then. With so much knowledge already at hand, why have these basic issues not been resolved?

This book shows that a single error in physics is preventing resolution of these issues. The error has to do with the classical notion of mass, and gravity, and especially the nature of Newton's gravitational constant G_N. For over eighty years, our understanding of General Relativity has been distorted by the persistent retention of an erroneous classical notion—that G_N is a fundamental constant. Once this erroneous notion is recognized and corrected, many of today's pressing problems in physics, including those mentioned above, become essentially trivial and are easily resolved using standard physics.

Earlier this century Einstein's General Relativity (GR) replaced the older Newtonian concept of gravity, but that replacement process remains unfinished. Using only simple logic, basic principles of physics, and the fundamental concepts of GR to complete the process gives answers to the questions posed above and provides substantial unification in physics, including the merging of QED and GR.

Aimed at bringing clarity and logic to the muddled mess now prevailing in fundamental physics, this book should be readily readable by both laymen and expert.

To quote Neil Armstrong, the first man on the Moon:

Just one Small Step for Man, One Giant Leap for Mankind.

Physics—The Big Problem—Solved!

1.
The Great Error in Physics

The greatest error in all physics, perhaps the greatest in the history of science, has been with us for a very long time, and even today its existence remains generally unrecognized. The consequences of this error have now reached enormous proportions and have pushed physics to the edge of a crisis. It is now time to get things right. The roots of the problem lie far in the past, an incredible 313 years ago in the infancy of physics. It began with a simple and very easily understandable mistake, but over the years between then and now, the error has remained uncorrected and the consequences have grown enormously. The original error was unavoidable, and, considering the relatively primitive state of physics at the time it was made, inconsequential. The error should have been corrected as physics advanced. It could perhaps have been corrected with the advent of General Relativity in 1917, or in the 1920s or 1930s, but it was not. Instead, the continuing failure to recognize the error even exists has caused the problem to be compounded, generating a series of errors, one built upon another, until now the result is a major crisis in physics that continues even today.

The error began with Isaac Newton (1687), continued with Cavendish (1798), and persisted through Einstein (1917). The impact on physics caused by this initial single error is now incredibly far-reaching. Although simple in origin, in no sense today can the error be considered small. Both philosophically and numerically, it is a very large error. It is now the primary cause preventing significant progress on the major problems facing physics today, and in particular is the direct cause of the failure to merge Quantum Electro Dynamics (QED), and General Relativity (GR), into a single comprehensive unified theory. The error concerns the nature of the property known as mass and the effect we call gravity, and thereby directly involves Newton's gravitational equation ($F_g = G_N m_1 m_2/R^2$), and in particular the classical gravitational constant, G_N. A major misunderstanding persists as to the nature of the property we call mass, and has led to a misunderstanding of the nature of G_N. And misunderstanding G_N has confused our interpretation of GR since it was introduced.

Physics today is largely built upon two main theories, Quantum Electro-Dynamics (QED) and Einstein's Relativity theories, especially General Relativity (GR). These theories are very successful and considerable experimental evidence supports their validity. However, physics has so far failed to merge these theories into a single Unified Theory—the search for which is now the "Holy Grail" of physics.

Although QED and GR have proven hugely successful in many different respects, there remain significant unresolved issues within fundamental physics. A partial list of these includes:

1. The merger of QED and GR into a single Unified Theory is necessary.
2. An explanation of the presence of large numbers of about 10^{40} within physics is required.
3. It must be explained why the conjectured Planck Scale* of matter is about 10^{21} from the naturally occurring stable particles, the electron and proton.
4. Understanding how the properties of mass and charge arise at the particle scale is necessary.
5. An explanation is needed of why gravity acts the same for all matter, i.e., why the same value of G_N evidently applies to both the electron and proton.
6. The mass asymmetry between the electron at 0.511

*The Planck Scale of matter is defined as that scale at which the fundamental forces of nature are equal and is also that at which the quantum fluctuations of the empty vacuum are believed to occur. The scale is named after Planck due to his conjecture that such a scale exists. Current numerical values for this scale are given later.

MeV/c^2, and the proton at 938.2 MeV/c^2 must be explained.

7. An understanding is required of how the fundamental forces, the Strong Force, the Electromagnetic force, and Gravity, arise and why they appear in the ratios 1: 1/137: ~10^{-42}.

There are of course other substantial issues within physics, but it is clear that the major problem is the fundamental nature of matter at the elementary particle scale is not understood. For some decades now, attempts have been made to resolve the general lack of understanding by "hurling mathematics at the problem," but with no success. String theory is one such example. Although the application of mathematics often brings clarity to a vaguely understood phenomenon, guessing at mathematical solutions seldom brings fundamental understanding. The major point on which all past efforts to merge QED and GR have failed is explaining the factor of approximately 10^{42} between electromagnetic effects (i.e., QED) and gravity (i.e., GR). No significant progress has been made on this issue in recent years, and as a result physics has now been essentially at a standstill since about 1970. In retrospect this is hardly surprising as the solution to this impasse is not to be found via new and complex mathematics or even a bizarre new theory, but rather by simple logical analysis of current basic physics. Careful analysis shows there are fundamental inconsistencies in physics regarding the notion of mass, and these inconsistencies are the root cause of the prevailing problems. Inconsistencies cannot be resolved by mathematics but only by the application of logic. Resolving these inconsistencies in a logical manner, consistent with the basic tenets of physics, removes most of the present difficulties and brings about considerable unification of many now disparate areas.

The difficulty with merging QED and GR as in issue #1 is that the electromagnetic (EM), and gravitational effects these two theories described have strengths apparently and inexplicably differing by about 10^{42}; the same numerical value as in issue #2. Issue #3 relates to the standard Planck Scale mass derivation that involves the root of G_N, (i.e. the Planck Mass, $M_P = \sqrt{[\hbar c/G_N]}$), so the factor of 10^{21} has to do with G_N and likely also derives from a factor of 10^{42}; (i.e. $\sqrt{10^{42}} = 10^{21}$). Issue #4 relates to how mass and charge effects emerge, with strengths differing by about 10^{42}, and issue #5 directly concerns G_N. Issue #6 has to do with the elementary particle masses, i.e. it relates to issue # 4, and issue #7 relates to the force ratios, again involving a factor of 10^{42}. Reviewing this list of issues, it may be surmised that if a reason for the presence of the large numbers of $\sim 10^{42}$ can be found, then each of the above issues can be resolved and a strong convergence of fundamental concepts will occur. This book shows this is in fact the case. As stated above, solving these problems does not require a new fundamental theory. All that is necessary is a logical review of current ideas, and the use of accepted basic concepts to squeeze out the unrecognized inconsistencies and muddled notions now prevailing in some areas of physics. Specifically, the basic concepts of GR must be used to obtain a clear understanding of the nature of mass and gravity and how they arise at the particle scale.

Our current understanding of gravity is based on Einstein's GR theory that describes it as an effect due to curvature of the metric of space. This description is not in question. The above list of unresolved problems relates to gravity, but does so via Newton's classical gravitational constant G_N, which is not obtainable by means of, and does not arise within GR. Einstein's GR is a mathematical theory and, unlike QED, is not inherently scaled to the observer's domain by the internal presence of any natural physical constants. Obviously GR and Newton's equation, and thereby G_N, must necessarily be somehow related as both describe gravity, although one (GR) is theoretically based and the other (Newton's) is derived empirically. Hence a specific connection between GR and Newton's equation has historically been presumed that preserves G_N as an elementary natural constant and thereby scales GR. *The preservation of G_N as a natural constant is the primary source of the impasse now existing in fundamental physics.*

The basic problem is that physics has not yet recognized all of the logical implications of General Relativity (GR), in relation to Newton's famous classical gravitational equation, $F_g = G_N m_1 m_2/R^2$. It is still widely believed within physics that Newton's classical equation is somehow valid, even though Einstein's GR theory shows that gravity is not a force but is a perceived attraction arising due to curvature of the metric of space. However, if GR is correct, and it appears that essentially it is, then Newton's equation is obviously fundamentally invalid as gravity, not being a force, cannot be described by a relation of the form: Force (F) = Function (anything). Other major aspects of Newton's relation can also be shown invalid by considering the fundamentals of GR.

Firstly: In physics, energy is defined as the ability to do work *in the observer's space-time domain* (i.e., the domain in which we exist), and Einstein's equation $E = mc^2$ shows mass (m) is c^2 removed from energy (E). Hence it logically follows mass is not a property that exists in the observer's domain, but a property that is necessarily c^2 distant from energy *and therefore also* c^2 *from the observer.* It is also well recognized in physics that multiplying a parameter by c, the velocity of light in empty space, implies a dimensional change. Any parameter defined in a space but moving at c in that space is no longer in that space, i.e., it has changed dimension.

For example, a three-dimensional cube in a frame of reference moving at velocity c to the observer is perceived by the observer as a two-dimensional plane. With energy defined in observer space, Einstein's equation shows that mass is a property twice removed from the observer by the velocity of light (c). Mass has to undergo a two-dimensional change to become energy: this is the essential mathematical meaning of $E = mc^2$ (see Appendix A). *It follows that gravity, defined as an attraction between particle masses, cannot operate in the observer's domain as implied by Newton's relation as the property of mass does not exist in the observer's domain, but in a domain* c^2 *distant.* It is important to note particles are individual entities and the mass of each particle is *independently* c^2 distant from the observer.

Secondly: It further follows that as Newton's equation is invalid (because it purports to describe an effect that does not exist in observer space), then the classical gravitational constant

G_N is a fictitious constant as it originates only from Newton's equation. Obviously, logic therefore dictates that because $E = mc^2$, the classical constant G_N cannot be a natural constant.

In Newton's equation, the apparent attraction of masses, and the classical G_N only relate to the observer's perception of the gravitational effect as experienced in his own domain. Neither Newton's equation nor the classical constant G_N relates directly to the real gravitational interaction that must operate between masses, with each mass being in its respective domain c^2 distant from the observer. If gravity operates between masses, and all masses are independently c^2 distant from the observer, then all masses are necessarily c^4 distant from each other. Consequently any relation (such as Newton's), attempting to describe the observer's perception of gravity as "a force acting between masses in the observer's domain" must include a $1/c^4$ term. Newton's relation does not include any such term.

As long recognized by physics (P.A.M. Dirac, 1928), all matter and specifically elementary particles are simply electromagnetic (EM) energy somehow spatially localized in observer space. From the observer's viewpoint, a mass held in the hand is clearly not energy (although it is equivalent to energy), and appears to exert a downward force due to the effect of gravity between the mass and the Earth. If the gravitational interaction between masses has the apparent dimensions of force, then the "force" must act in the c^2 distant domains in which the masses exist. In other words, if gravity is equivalent to a force acting between masses, the force acting on a mass in the c^2 distant domains of the particles, is $F_g = G_S m^2/R^2$, where G_S is a constant of the interaction in the c^2 distant frames. The attraction (A, say), evident to the observer will appear to be $1/c^4$ less and equal to $F_g/c^4 = A = (1/c^4)G_S m^2/R^2$. This apparent attraction is clearly the effect represented by Newton's equation, wherein the classical constant G_N is represented by $(1/c^4)G_S$. This raises the question of how an apparent constant, even a fictitious one such as G_N, can contain a term such as $1/c^4$ whose value is dependent on the units of c. In other words, how can a variable such as $(1/c^4)G_S$ be consistent with an empirically determined "constant" such as G_N? The manner by which the $1/c^4$ term in G_N escaped the attention of physicists is perhaps one of the key reasons why the error, i.e., the erroneous belief that G_N is a natural constant, persisted unrecognized for so long. The reason is again simple, quite logical, and is as follows.

If the real gravitational interaction operating in the c^2 distant domains is essentially equivalent to a force, given by $F_S = G_S m^2/r^2$ say, then G_S must have the dimensions currently assigned to the classical constant G_N. These dimensions are those of $\hbar c/m^2$, where $\hbar$ is $h/2\pi$ and h is Planck's constant. If this "equivalent force" is first described in one particular system of measurement units, say cm.gm.sec. (cgs) units, and then converted into another, SI units (say), G_S converts numerically as does $\hbar c/m^2$, and F_S converts as $\hbar c/r^2$; [the units of F_S are $G_S m^2/R^2$, so as $G_S \sim \hbar c/r^2$, $F_S \sim \hbar c/r^2$]. Consider now the similar conversion of the apparent attraction evident to the observer and described by Newton's erroneous equation, now represented by A. The apparent attraction as measured in SI units will differ from that measured in cgs units by the same factor, as does the real effect, i.e., as by conversion of the units of $G_S m^2/R^2$. Hence, *with* $A = (1/c^4)G_S m^2/R^2$, *experimental consistency is only obtained on converting from cgs to SI units by maintaining the* $1/c^4$ *term as a constant with a value given by the units in which it was first determined. As the presence of* $1/c^4$ *is not suspected in A, maintaining it as a fixed numerical value will achieve apparent consistency of measurement between all systems of units.*

Hence incorrectly describing gravity as a force between masses in observer space has two consequences:

1. The $1/c^4$ term is inadvertently included within G_N and is hidden therein.
2. On converting between measurement systems, experimental consistency requires that the hidden $1/c^4$ term in G_N masquerade as a constant in the measurement units in which it was originally measured; i.e., in the cgs units used by Cavendish in 1798.

These errors have inadvertently introduced a factor of c^4 in cgs units (~8.08 x 10^{41}) into G_N and thereby into all areas of physics involving G_N. The consequences of this error are now huge. For example, the basis of much current theoretical physics research, the Planck scale mass M_P, is given by $M_P = \sqrt{(\hbar c/G_N)}$. Clearly, as G_N contains an error of ~10^{42}, the Planck scale mass is in error and is too large by ~10^{21}. Not surprisingly, as shown later, correcting this error brings the Planck scale into correspondence with the natural particle masses, thus resolving a major problem in physics, and issues #2 and #3 as listed above.

This then is the series of errors that has led physics astray:

— Newton's assumption that mass exists in observer space is incorrect, and that error led to his famous equation, now seen to be invalid.
— This invalid equation introduces G_N, a classical constant now seen to be fictitious.
— Measurement of G_N by Cavendish in cgs units inadvertently introduced a large number of about 10^{42} into physics.

It is clear that $G_N \sim G_S/\underline{c}^4$, where $\underline{c}$ is used to signify that c is in cgs units, and $G_S \sim 10^{42}\ G_N$. This is also the origin of the fundamental force ratios mentioned in issue #7, as the true "force of gravity" acts between particle domains that are c^4 apart and at the strong force scale within those domains. Resolution of the remaining issues quickly follows.

There is a larger aspect to the error in presuming Newton's relation and G_N are both somehow valid. Einstein's General Relativity (GR) theory describes gravity as due to metric curvature and is presumed essentially correct, but the theory does not contain any internal scaling parameter, consequently GR has been scaled by using G_N. This means that GR has heretofore been viewed as a cosmological scale theory, whereas as is now evident (from the error in G_N), it is not. Elementary particles are often described as the metric of space "curled up into a sphere of microscopic dimensions," and it is now clear that *GR is a theory directly relating to metric curvature on the microscopic elemen-*

tary particle scale via the "strong force scale" constant G_S. Hence on this scale both QED and the re-scaled GR both directly apply to elementary particles. The erroneous impression that gravity operates directly in the observer's domain, and therefore G_N scales GR, is only due to the observer's perception.

2.

Why Did the Problem Persist Unrecognized for So Long?

It was not until the twentieth century that the first opportunity to correct Newton's error arose. In 1917 Einstein published his famous General Relativity theory, showing that gravity is not a force that acts on masses at a distance, but an effect due to the curvature of the metric of space. Einstein thus showed that Newton's equation was fundamentally invalid, but for some reason the manner in which it was invalid was not effectively pursued. The greater implications of $E = mc^2$ should have been considered, as its meaning is clearly obvious—mass is dimensionally c^2 distant from energy, and thus dimensionally c^2 from the observer's domain. But this clear consequence of $E = mc^2$ seems never to have been raised.

General Relativity replaced Newton's concept of gravity but had to be reconciled with the experimentally useful Newtonian relation. Even though Einstein's theory was accepted as correct, and Newton's force at a distance concept was now known to be wrong, "everybody" also knew that Newton's equation worked, so "it must be valid somehow," and so the two had to be reconciled. This is where things went badly wrong due to a great "Double-Think." Apparent compatibility was achieved by forcing a "merger" whereby Newton's equation was described (i.e., "derived"), as an approximation to Einstein's General Relativity, even though the two concepts—gravity as due to a force and gravity due to metric curvature—are totally incompatible. The manner by which this "fit" was achieved was an error, a Big Error. This erroneous "derivation" is still to be found in physics books, even though it has been over eighty years since Einstein showed that gravity is not a force and hence Newton's equation is inherently invalid. This apparent reconciliation of Newton's classical equation with GR was a big disservice to physics as it masked the underlying problem, essentially removed motivation for further effort on the subject, and thereby derailed the progress toward convergence in fundamental physics. A relationship between GR and Newton of course exists, but it is a described herein and must not be a "forced derivation" based on preserving G_N as a natural constant.

Even though GR shows that gravity is not a force, for some hard to understand reason (outside of muddled thinking!), even today Newton's equation is still largely considered as essentially valid. Consequently, and even worse, mainstream physics still considers G_N as a fundamental natural constant! Evidently the old mechanistic ideas die very slowly. Erroneously presuming G_N is a fundamental constant led to an even larger error, as it is the basis by which GR is scaled and the basis by which the conjectured Planck Scale of nature is derived. The Planck Scale is defined as the scale at which the fundamental forces of nature are equal and the scale at which the quantum fluctuations of the empty vacuum are believed to exist. Hence getting the real gravitational constant wrong by about 10^{42} has more than a minor impact on our understanding nature!

Perhaps a major reason the error has been unrecognized for so long is that even though GR was introduced in 1917, physics was very mechanically oriented until the arrival of Quantum Theory in the 1920s and 1930s, and even then the old ideas were not instantly replaced. The mechanistic viewpoint has only slowly given way to modern ideas, and even today Newton's classical gravity relation has not yet been fully discarded. Indeed, except for Newton's relation and G_N there is no other means of calculating the effect of gravity, GR notwithstanding.

Perhaps of even more critical importance is that the physics community is very conservative. No longer is it composed of scientists simply seeking knowledge for its own sake, but instead it is a community that must now work for a living. Which physicist, striving for academic tenure, the approval of his peers, or a research grant is going to risk questioning Newton's famous equation? Newton—one of the great gods of science—wrong? Unthinkable! To compound the problem, try telling anybody, physicist or layman, that even though Einstein got the math right in his General Relativity theory, he misunderstood the physical aspects and thought GR was a cosmological theory, instead of the particle scale theory it is now seen to be. Einstein and Newton both in error!! Incredibly unthinkable! Even though scientists consider themselves open-minded and objective, this conservatism represents a huge psychological barrier to accepting the ideas presented here.

Another problem is that it is essentially impossible for those not in academia or the research community to get a paper published in any worthwhile technical journal, as peer review is a major hurdle. Academic tenure and status in the community evidently relate to the number of papers published, so the tendency is toward numerous papers with incremental changes. Large and sudden changes proposed by unknown authors are not well received. Papers submitted in recent years on this subject (by the author), have all been rejected with irritatingly inappropriate comments. Such as: "does not agree with accepted theory!" (Well, of course not, that's the whole point!) Or, "must have your professor's approval" (Huh?). "Please provide recommendations of three of your co-workers in the field," (co-workers?). Even worse, some reviewers have responded with: "everybody knows that G_N is a fundamental natural constant, and the numerical correspondence of the suggested factor $\sim 10^{42}$ in G_N with c^4 in cgs units must be just a coincidence." (Right, ignore the logic, call it a coincidence, to three significant figures over 42 orders of magnitude! See Appendix B.)

Einstein's General Relativity (GR) theory is sufficient to solve the problem of fully understanding the nature of gravity, and how it arises at the particle scale. But for whatever reasons, even though it is very apparent Newton's and Einstein's theories are fundamentally incompatible, the problem simply has not been given due recognition. Physics has "hand waved" and glossed over the issue, and the "official" viewpoint is that things are just fine and there is no big issue on this topic. This is obviously wrong.

The logic given above by which the problem with physics has been resolved is so simple and so compelling it is hard to understand why the obvious implications of Einstein's $E = mc^2$ have been ignored for so long. This is particularly surprising as $E = mc^2$ has become highly visible in popular scientific culture. Einstein showed that Newton's equation was fundamentally invalid, but the question of why it appears to work in practice has not been resolved in a satisfactory manner until now. Perhaps it is not polite to question why Einstein himself did not point out the problem and solve the issue, or why he did not point out that as gravity is not a force, G_N is not a fundamental constant. Then there is also the question of why nobody else has pursued and resolved these issues in the decades that have followed. But these questions are now moot.

3.
The Classical Gravitational Constant G_N

How is the problem of reconciling Newton's classical concept of gravity with Einstein's GR resolved? The answer is simple and based on the above is as follows: (see Figure 1).

1. Each mass exists in a frame of reference distant c^2 from the observer and consequently c^4 from all other masses.
2. The gravitational attraction acting on masses in these c^2 distant domains can be described by an equivalent force in those domains, $F = G_S m_1 m_2/R^2$, where G_S is not Newton's gravitational constant. As above, it can be easily shown G_S is a constant of Super Gravity that operates at the Strong Force scale, some 42 orders of magnitude stronger than observer gravity.
3. As masses are c^4 distant from each other, the attraction evident to the classical observer in his own space is $A = (1/c^4)F \sim (1/c^4)\, G_S m_1 m_2/R^2$, where $(1/c^4)G_S$ is numerically equal to Newton's classical gravitational constant G_N. How a "constant" can have a value dependent on the units of $1/c^4$ has already been shown.
4. With $G_N \sim (1/\underline{c}^4)G_S$, where $\underline{c}$ signifies c is always in cgs units, and $G_S = \hbar c/M^2$, a value for M can be determined by numerical comparison. The value so determined is 26.58 m_e, i.e., 26.58 times the electron rest mass in observer space. As shown later, this factor of 26.58 arises directly from consideration of the electron as being rotationally relativistic and relates to the dimensionless Fine Structure Constant alpha (α), of

atomic physics, where $e^2 = \alpha\hbar c$, (with e the electron charge), as given by Sommerfeld. Alpha has a numerical value of 1/137.03 and $\alpha^{2/3} = 26.58$, so $M = m_e/\alpha^{2/3}$.

As shown later, an expression for G_N can be derived in terms of the atomic constants. However, it is emphasized that G_N is not a natural constant, but relates only to the observer's perception. The expression by which the numerical value of G_N can be calculated in any units system is:

$$G_N \equiv (1/\underline{c}^4)\,\hbar c/(m_e/\alpha^{2/3})^2 \tag{3–1}$$

where $\hbar$ is $h/2\pi$, and $\underline{c}$ shows c is always in cgs units and m_e is the electron mass. A numerical value for G_N based on this relation and the standard values of the atomic constants is given in Appendix B.

The error of including c^4 in cgs units in G_N has largely decoyed basic physics "off track" for over eighty years and has had monumental consequences: These include:

a. The basic strengths of the fundamental forces, the "Strong Force" and gravity, are thought to be about 10^{42} apart, whereas as is now evident, gravity in the observer's frame is a direct consequence of Super Gravity acting at the strong force level within particles. Gravity is simply the residue of the Super Gravity induced metric curvature in the particle frame rotating at essentially c, that is evident in the observer's frame.
b. The currently accepted Planck Scale of nature, established by equating the fundamental forces, is in error by 10^{21} and when corrected coincides with the naturally occurring particles.
c. The big problem in current physics, establishing compatibility between GR and QED, is now easily resolved via a conceptual model for the electron, as shown below.
d. The natural particles constitute space rolled up into small spheres of the order of 10^{-15}m, not at the currently conjectured Planck scale, ***showing that GR is a the-***

ory scaled to elementary particles, and is not a theory on the cosmological scale.

e. A simple electron model is readily derivable, showing it to be a harmonic oscillator, composed simply of spatially localized electromagnetic energy in dynamic equilibrium between electrostatic repulsion and Super Gravitational self-attraction. The concept is also describable as a relativistic mass circulating in a precessing elliptical orbit, and thereby quantized in angular momentum. As explained later, the relativistic state only exists relative to the observer, within the local particle space the energy is moving at the local velocity of light in a curved metric. The local velocity of light differs from c, the free space value due to the presence of the local particle energy.

f. The vacuum mass density due to quantum fluctuations is now calculable and, as shown later, is close to that observed in the cosmos.

g. String Theory is easily reconciled with the concepts of Super Gravity and mass existing in c^2 distant domains, i.e., the fundamental gravitational interaction occurs at the strong force level in the c^2 distant frames.

h. Confusion has apparently continued to exist over the basic nature of matter for some time. No longer, it is simply spatially localized electromagnetic energy as stated by Dirac in 1928.

i. It is easily established how mass (i.e., gravity) and electric charge arise at the elementary particle scale by EM energy localization. Energy localization causes formation of an equilibrium state where two metric stresses are in balance. One metric stress is radial and its residual effect seen in the observer's domain is perceived as electric charge; the other is circumferential and causes a spherical metric deformation at effectively the strong force scale in the particle's rotating frame. The residue of this metric curvature is seen as gravity by the observer.

j. etc. etc.

The writer does not expect these assertions to be accepted at face value, so these claims are substantiated by logical arguments based only on accepted physics that follow in the next section.

Figure 1

A: Newton's erroneous concept of gravitational attraction as a force acting between static masses existing in observer space, giving $F = G_N m_1 m_2/R^2$.

Compared to:

B: A quasi-classical representation of elementary particles as mass rotating effectively at velocity c, and existing in the observer's domain as spatially localized energy. This gives an apparent attraction $A = (1/c^4)\, G_1 m_1 m_2/R^2$.

A

B

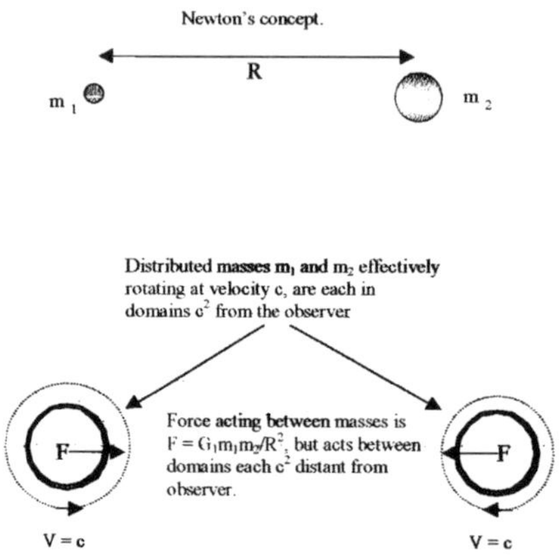

The classical observer sees mass attraction $A = F/c^4 = (1/c^4)G_1 m^2/R^2$, and correctly interprets it as force acting on masses, but erroneously believes it acts in his domain. This leads to $G_N \sim (1/c^4)G_1$.

4.
Reconciling GR and QED

The compatibility problem between QED and GR can be resolved by considering only the simplest concepts of modern physics. Logical considerations based on $E = mc^2$ show the 10^{42} separation of the two theories is not real and is but an accident of history.

Consider a local metric distortion, static in the observer's frame of reference, interpreted as a mass by the classical observer and typically represented by the symbol M. A similar metric distortion, moving in a plane at velocity c, the velocity of light in free space relative to the classical observer, is interpreted by the observer as momentum, $p = Mc$, in that plane. If that plane were then to be translated at a velocity c along an axis normal to the plane, the observer would perceive the metric distortion as energy, given by $E = pc = Mc^2$. This relation shows that energy, in principle, is mass moving at velocity c in two directions simultaneously. (See Appendix A.)

Consider now a metric distortion as described above that constitutes a mass moving at velocity c in two dimensions in a spherical shell such that to the observer it would appear as a spatially localized energy E. Classically, the force required to maintain a mass M in a spherical orbit at velocity v and radius R is $F = Mv^2/R$. This force acts on the mass and therefore exists in the rotating frame of the mass, not in the observer's frame. For a mass moving at velocity c, the force acting on the mass is $F = Mc^2/R$, where the mass M is that in the rotating frame. If such a system were to exist, in the observer's frame, both the apparent mass and the orbit radius would be reduced to zero by relativistic effects, although a local energy would exist. Such a system would hence represent infinite energy density in the observer's frame and is therefore impossible: it is called a singularity.

Planck's relation $E = \hbar c/r$, gives the energy of a single electro-magnetic (EM) photon of wavelength $2\pi r$, where $\hbar$ is the reduced form of Planck's constant h and $\hbar = h/2\pi$. In the circumstance where the energy exists in an orbit of radius R in the rotating frame, equal to r the energy wavelength/2π, the energy wavelength is equal to the orbit circumference and the orbit is a quantum orbit. A photon in an orbit is equivalent to a mass circulating at velocity c. In such a quantum orbit, the force acting on the mass is therefore $F = Mc^2/r$. Hence as $E = Mc^2 = \hbar c/r$, the force is $F = \hbar c/r^2 = F_S$, (say). In particle physics, a force of this strength is at the Strong Force scale, and the mass M circulating in a quantum shell at c as described is therefore constrained at the "Strong Force" level.

In the quantum shell as described, there is no entity other than the system mass and the localizing effect at the "Strong Force" scale can therefore only be represented by mass self-attraction acting in the rotating frame on the mass M. By inserting M^2, $F_S = \hbar c/r^2 = (\hbar c/M^2)M^2/r^2$. Such a mass attraction is by definition gravitational in nature but is far stronger than that evident in the observer's domain. If the term $\hbar c/M^2$ is taken as a Super Gravity force constant G_S, *not* equal to Newton's classical constant G_N, then $F_S = G_S M^2/r^2$.

Now $E = Mc^2 = \hbar c/r$, which on rearranging gives $r = \hbar c/Mc^2 = (\hbar c/M^2)M/c^2 = G_S M/c^2$. In the circumstance where M is composed of two similar masses each of M_q, we obtain $r = 2G_S M_q/c^2$. *This is exactly the form of the expression for the Schwarzchild radius (or event horizon) of a Black Hole of mass M_q as obtained from General Relativity.* If general relativity applies at the strong force scale and effectively operates with a quasi-classical force constant G_S, then the strong force is a form of Super Gravity and the system constitutes a microscopic Black Hole. In such a system, the local metric is highly curved to such a degree as to effectively be curled up into a ball, and the EM propagation is propagating rectilinearly in the local space. The local metric curvature constitutes the spatially constraining effect, and no actual force F_S need exist. To a distant observer unaware of the locally curled-up metric, the system energy appears spatially constrained as though by a force acting at the strong force scale, even though no such force exists.

Energy is defined as the ability to do work in the observer's domain. Therefore, as $E = mc^2$, the mass associated with an entity as described above does not exist in the observer's domain but in a rotating domain distant from the observer by c^2. This is the precise mathematical meaning of Einstein's relation $E = Mc^2$. Two similar particles existing in observer space, with the mass of each c^2 distant from the observer, will have masses distant from each other by c^4. The same mass attraction presumably operates between these distant masses as on mass in the same system, as the metric curvature arising from a localized mass only diminishes with distance, and can never terminate. Hence all masses are located in the curved metric fields of each other.

It is clear that for spatial continuity in three-dimensional space the metric distortion due to a microscopic black hole decreases with distance from the system as $1/d^2$. The observer therefore notes an apparent attraction between two mass systems of, $A \sim (1/c^4)G_S(m/d)^2$, where m is the apparent mass of the system to the observer. This apparent mass would be zero if the masses were truly rotating at c, the velocity of light in free space of unity metric (optical) index, i.e., $v = c/n$ where $n = 1$. In the metric region occupied by each system, space is not "free," and the presence of localized energy necessarily modifies the metric index, causing it to diverge from unity, i.e., $n \neq 1$. The EM propagation, self-trapped by metric curvature, although propagating at the velocity of light in the local metric, is no longer at velocity c relative to the observer. The observer therefore necessarily perceives the presence of a local mass, i.e., the metric distortion caused by the mass of the black hole is evident to the observer by virtue of its interaction with other masses, even though the mass is nominally c^2 distant from the observer's domain.

With the metric index in which the mass exists differing from unity, the mass M in the rotating frame will appear to the observer to possess a different mass value (m), by virtue of its relativistic state relative to the observer. For slower rotational states, the apparent mass will be higher, in the limit reaching M for a rotational velocity of zero. As the mass is propagating rectilinearly in a curved metric, the Lorentz relations of special relativity apply and the effective mass in the observer's domain can be represented as $m = \beta M$, where $\beta = 1 - (v/c)^2$, and where v is the rotational velocity of the energy relative to the observer. It is interesting to note the wavelength associated with the mass energy apparent in the observer's domain will be larger than the measurable system circumference of h/Mc.

If the total metric volumetric distortion due to a mass M is proportional to M ($= m/\beta$) in the rotating frame, then by symmetry considerations, the distortion (contraction or expansion) per spatial dimension is proportional to $m/\beta^{1/3}$ ($= \beta^{1/3}M$). The metric distortion affecting a distant mass is a gradient and is proportional to $M\beta^{1/3}$. Hence two similar systems rotating at $v = c(1-\beta)^{1/2}$, distant from each other by d in the observer's domain, cause an apparent attraction evident to the observer of $A = (1/c^4)G_1(m/d)^2$, where $G_1 = \hbar c/(M\beta^{1/3})^2 = \hbar c/(m/\beta^{2/3})^2$.

This apparent attraction of masses, $A = (1/c^4)G_1(m/d)^2$, has been historically mistaken for a force acting between masses in the observer's frame. This error, persisting since 1687, led to the Newtonian gravitational relation $F_g = G_N m_1 m_2/R^2$, which is now seen to be invalid. As $E = mc^2$, it is clear that mass is not a property in the observer's domain and consequently, although gravity acts between masses, the force of gravity does not act in observer space but between masses each c^2 from the observer. This has several consequences:

1. The mass interaction operates in the c^2 distant frames of particles in which mass exists. The mass interaction in those frames is at the strong force scale, and only appears c^4 weaker to the observer. *Hence the original string theory of Veneziano (1965) is essentially correct.* Gravity can be considered as an interaction due to the transfer of a spin 2 graviton between particles that occurs at the strong force scale. The interaction occurs in the frames rotating at c and thereby c^2 distant from the observer.

2. The classical observer incorrectly interprets the attraction A as a force in his domain and thereby derives a gravitational force constant G_N. It is clear Newton's equation is invalid, as such a force does not exist, and there is no natural constant such as G_N. The numerical value first established for G_N by Cavendish (1798) is equivalent to $G_N = (1/c^4)G_1 = (1/c^4)\ \hbar c/(m/\beta^{2/3})^2$. This first value was obtained using the cgs system of units. Numerical comparison of this expression in cgs units with m the electron mass against the empirical value of G_N gives a value for $1/\beta^{2/3}$ of 26.58, indicating $1/\beta$ is 137.03. This is the same value as the inverse fine structure constant $1/\alpha$.
3. The true localizing influence of gravity, i.e., metric curvature, acts within the particle rotating frame and is equivalent to a force with a "force constant" of $\hbar c/m^2$, leading to a force equivalent $F = (\hbar c/M^2)M^2/r^2$ in that frame. On changing from a cgs system of units to any other units system, this equivalent force changes per these dimensions, as does A in the observer's domain. The remaining $(1/c^4)$ term hidden within A is unaffected and therefore acts as a constant of value $1/c^4$ in cgs units, i.e., a value of 8.078×10^{41}. This numerical value has therefore inadvertently become included in the classical value for G_N, and thereby appears in physics everywhere that G_N occurs.

   For example, the value of the currently accepted ratio "electromagnetic force/gravity" for the electron is $\alpha\hbar c/G_N m^2$, or 4.16×10^{42}, which is exactly $\alpha^{-1/3}c^4$ in cgs units.
4. With $\beta = \alpha = 1/137$ the nominal mass in the rotating frame is 137 m = M = 70 MeV. The value M_q is 35 MeV. The muon and pion masses at 106 and 141 MeV respectively indicate a 2:3:4 sequence with a mass increment of about 35.3 MeV. Interestingly, the proton mass at 938.3 MeV divided by 35.3 is again 26.58 (as in #2 above)! More on this later.
5. The presumed Planck Scale of matter, that scale at

which the fundamental forces of nature, strong, electromagnetic, and gravity, are equal, is now seen to be misrepresented by equating the strong force in the c^2 distant rotating frame with the effect of gravity in the observer's domain. It is more appropriate to equate the "strong force" with Super Gravity in the rotating frame, giving $\hbar c/r^2 = G_S(M/r)^2$, or $M = (\hbar c/G_S)^{1/2}$. This is clearly trivial as by definition $G_S = \hbar c/M^2$. If the above system is identified with the electron with $M = m_e/\alpha$, then the electron is seen to be a Planck scale particle - with internal forces equal. It can be envisioned as a harmonic oscillator, in dynamic equilibrium between Super Gravity mass attraction and electrostatic self-repulsion at the strong force scale in the rotating frame.

6. Localizing an energy Mc^2 about a radius r from r = infinity creates a local kinetic energy of Mc^2/r where the energy wavelength is $2\pi r$. A compensating potential is necessarily created by reason of energy conservation with a value of $\hbar c/r^2$. The system so formed is rotationally relativistic and in the observer's frame, both the mass and potential appear reduced by a factor of alpha ($\alpha \sim 1/137$) to a mass m and a co-located electric potential energy $\alpha\hbar c/r^2$, or e^2/r^2 as $e^2 = \alpha\hbar c$ by Sommerfeld. Hence energy localization necessarily forms a local mass and a unit charge, causing the local metric to be "non-free."

It is interesting to speculate that as particles and anti-particles have opposite charge values, the change induced in the local metric is also opposite in nature. Thus whereas some particles may constitute energy rotating at slightly less than c relative to an observer, the corresponding anti-particle may be rotating at slightly above c, relative to the observer. This also leads to speculation as to the nature of protons versus electrons. For example, suppose the electron is rotating at less than c, then the positron may be rotating at greater than c and would not be well suited to existing in this universe. The proton may be an "inside out" black hole where, also being positively charged, causes a double symmetry inversion that leads to a state where the energy is also rotating at less than c. Hence the electron and proton and their anti-particles may represent a four-way symmetry.

Conclusion: The means of achieving the much sought after reconciliation between QED and GR is now obvious. It is achieved by re-scaling GR to a particle scale theory operating at the strong force level. Current perceptions of GR as a cosmological scale theory are in error and are due to the misunderstanding as to the nature of G_N.

5.
The QED Estimate of the Vacuum Energy Density

One of the most perplexing issues within physics is that the current estimate of the vacuum energy density derived by Quantum Electro-Dynamics (QED), is about 10^{122} greater than is supported by direct cosmological observations. This issue is readily resolved by fundamental considerations that obtain a value for the vacuum energy density in close correspondence with observations.

Quantum fluctuations are believed to continuously occur at the Planck scale of matter whereby electric charges spontaneously appear and disappear, resulting in a non-zero average energy level in the otherwise empty vacuum. In the "Standard Model," this vacuum energy is estimated by using the "Planck Scale" parameters. These are derived via Planck's notion (circa 1899) that "a scale of matter exists at which all the forces of nature must be equal." Based on this notion, the currently accepted derivation equates the strong force ($F_S = \hbar c/r^2$) and Newtonian gravity ($F_g = G_N m_1 m_2/r^2$), as in (5–1), leading to the Planck Mass $M = (\hbar c/G_N)^{1/2}$, at 1.221×10^{19} GeV/c^2.

$$\hbar c/r^2 = G_N M^2/r^2 \qquad (5\text{-}1)$$

(G_N is Newton's gravitational constant, $\hbar$ is the reduced Planck constant - $h/2\pi$, c is the velocity of light in free space). The Planck Mass relates to a length r, via $E = \hbar c/r = Mc^2$, and thereby allows calculation of the Planck Scale mass density of 1.23×10^{93} grams/cc. due to a mass M in a sphere of radius r.

Cosmological observations indicate the universe is essentially "flat" with a cosmological constant of less than $(1 \times 10^{-26}$ meters$)^2$, and place the average mass density in the universe from all sources, matter and vacuum fluctuations, at about 0.9×10^{-29} grams/cc as shown in **Figure 2.** Vacuum energy fluctuations are expected to contribute about half of this value, showing the QED estimate of the universal mass density of 1.23×10^{93} grams/cc is in error and is too large by approximately 2.7×10^{122}. The ratio between the Planck and cosmological mass densities gives the ratio between the corresponding energy densities. By this calculation the energy density due to vacuum fluctuations at the Planck scale is nominally 10^{122} greater than the universal vacuum energy density estimated from cosmological observations.

This huge discrepancy between theory and observation can be resolved as follows. Consider an electromagnetic wave of energy $E = \hbar c/r$ of equivalent mass energy mc^2, forming a quantum orbit of radius r: (singularity issues are not addressed here—see Note 1). The energy E is referenced to the observer's domain and in principle is equivalent to a mass m rotating at velocity c about a radius r, i.e., forming a system of one wavelength circumference. Within a coordinate frame rotating at c, the system constitutes only mass. From the observer's perspective, the system energy is apparently localized by a force, which from a simplistic classical viewpoint necessarily acts via the mass to maintain such a system. The apparent force (Note 2), logically considered by the observer is therefore $F = mc^2/r$, which as $mc^2 = \hbar c/r$ is $F = \hbar c/r^2$, is hence at the Strong Force level. As the "apparent force" acts via the mass, by expansion it can be written as an inverse square mass attraction in the rotating frame, hence:

$$F = (\hbar c/m^2)m^2/r^2 = G_S m^2/r^2 \qquad (5\text{-}2)$$

where $\hbar c/m^2$, ($= G_S$, say), can be taken as a "force constant."

For a spatially localized particle, the "apparent force" is considered as acting in three dimensions over a spherically distributed mass rotating everywhere at velocity c. Therefore, by $E = mc^2$, the mass is dimensionally distant from the observer's domain by c^2. G_S is effectively a "Super Gravity" force constant and if m is taken as the electron mass, G_S (= $\hbar c/m^2$,), is about 40 orders of magnitude greater than Newton's G_N.

Consider two such systems distant from each other and interacting gravitationally via metric fluctuations propagating at velocity c in free space (i.e., gravitons, the usual gravity force exchange particle). The mass interaction at each system location occurs between a graviton moving at velocity c and a mass rotational at c. This interaction is effectively between two masses in a frame rotating at c and therefore, from the observer's perspective, occurs at the strong force level as in Veneziano's original (1968) String Theory. The Super Gravity "apparent force," F, acting within particles, is due to a metric distortion as per General Relativity (Note 2). In essence, at the elementary particle scale the space time metric is "rolled up" into a microscopic sphere to form additional dimensions. The surface of a sphere rotating everywhere at velocity c provides at least two additional spatial and two additional time dimensions, each separate from the observer by c.

To the observer the equivalent force acting in the rotating system appears reduced for two reasons;

1. the "force" F, acts in a frame rotating at c and hence in a domain dimensionally distant from the observer by c^2.
2. metric distortion at the distant location of a second mass due to the first is effectively two dimensional, compared to the three dimensional spherically distributed mass distortion within a system.

The attraction between two distant similar masses, each rotating in separate c^2 distant frames, as described above is evident to the observer as an apparent attraction A, as in (5-3). The effective mass of each, by reason of 2) above, is not that internal to the particle (i.e., m), but is modified (Note 3) to m_1.

$$A = (1/c^4)G_S m_1^2/r^2 \qquad (5\text{-}3)$$

Attempting to determine a "Planck Mass" by equating the "gravitational attraction" A, experienced in the observer's domain, with the strong force ($\hbar c/r^2$) in the rotating frame introduces $1/c^4$ into the "equality," and gives an invalid expression of the form "$m = c^2(\hbar c/G_S)^{1/2}$", whereas $m = (\hbar c/G_S)^{1/2}$ by the above definition of G_S. If this erroneously presumed "equality" is used to obtain a "Planck Mass," the derived value has dimensions of mc^2, effectively assigning the estimated "Planck Mass" too large a value by c^2. The corresponding Planck energy is similarly too large by c^2.

Consider that the vacuum energy density is not due to the presence of Planck Scale masses everywhere, i.e., a fluctuation in three dimensions, but due to quantum electromagnetic fluctuations in the metric of the vacuum occurring independently in each dimension. Consequently we must consider not the "Planck Mass per unit volume" but the corrected energy per unit length. With the classical "Planck Energy" too large by c^2, the corresponding "Planck Length" is too small by c^2, and the resulting "Planck Energy per unit Length" is too large by c^4. As the vacuum volumetric energy density is the cube of the vacuum energy density per unit length, the vacuum energy density estimated via the currently accepted (QED based), Planck Energy is too large by $(c^4)^3$, or about c^{12}.

Returning to the attraction between masses that is evident in observer space, the relation $A = (1/c^4)G_S m^2/r^2$, clearly corresponds to the classical gravitational relation $F_g = G_N m^2/r^2$, now seen to be invalid as true gravity operates on mass within systems rotating at c and is due to metric distortion, not a force in the observer's domain. Newton's gravitational constant G_N, measured first by Cavendish in 1798, correlates to $(1/c^4)G_S$, with the value of m in G_S modified for the reason given in 2) above. Newton's constant G_N therefore erroneously contains a factor of c^4 in the units in which Cavendish made the measurement – c.g.s. units. The real gravitational interaction acts within the rotating frames, consequently, on converting the classical gravitational expression into another system of units (e.g., MKS), the force constant G_S changes as $\hbar c/m^2$ and the c^4 term within G_N behaves as a fixed value. These historical errors (omission of c^4 from G_N, and gravity being considered a force in observer space), have introduced a factor of c^4 in cgs units into all parts of physics involving G_N. *The "Standard Model" Planck scale is hence in error by* c^2 *in cgs units, a factor of 8.99 x* 10^{20}*, and the QED estimated vacuum mass density is in error, and is too large by nominally (2.998 x* $10^{10})^{12}$ *or about 5.3 x* 10^{125}*, pending other smaller adjustments.* Correcting the current QED estimate of 2.7 x 10^{122} by 5.3 x 10^{125} leaves an unresolved error of 2.7 x 10^{122} / 5.3 x 10^{125}, a factor of about 1/2,000. This residual difference requires some adjustments as below.

Adjustments:

a. The current "Planck Mass" is seen to be c^2 too large and should be corrected to approximately (1.221x10^{19} GeV/c^2) / 8.99x10^{20} = 13.6 MeV/c^2, which at 26.6 electron masses is nominally the same order of magnitude as the naturally occurring particles. As explained in Note 3, a more appropriate "Planck Mass" is half the mass in the electron's relativistic rotating frame, $m_e/2\alpha$, or 137/2 times the electron rest mass evident to the observer. From Note 2, the "Planck Mass" implied by Newton's G_N is $m_e\underline{c}2/\alpha^{2/3}$, as $G_N = (1/\underline{c}^4)\,\hbar c/m_1^{\,2}$ (where $m_1 = m_e/\alpha^{2/3}$, and $\underline{c}$ indicates c is always in cgs units). Hence the current Planck Mass is too large by $2\underline{c}^2\alpha^{1/3}$, compared to the new "Planck Mass" (internal to the electron), of $m_e/2\alpha$. This modifies the vacuum energy density correction factor from c^{12} to $(2\underline{c}^2\alpha^{1/3})^6$ or $(2(2.998\ \text{x}10^{10})^2 / 5.155)^6$ or 1.8 x10^{123}, and provides a partially corrected vacuum mass density estimate of 1.23 x10^{93} / 1.8 x10^{123} = 0.68 x10^{-30} grams/cc.

b. The value 0.68 x10^{-30} grams/cc., is the estimated vacuum mass density in the rotating relativistic frames due to half the electron mass. The electron vacuum energy density in the observer's domain would therefore be 137/2 times less at 9.93 x 10^{-33} grams/cc.

c. The vacuum mass density is due to the quantum fluctuation of electric charge pairs in which electrons and protons presumably participate in equal numbers, both for reason of charge conservation and as they appear equally abundant in the universe. The proton mass is greater than the electron mass by a factor of 1836. If vacuum fluctuations are equally distributed between both particle types, the implied fluctuation mass density would be increased by 1836/2 to 0.91 x 10^{-29} grams per cc. However, it is not clear both the electron and proton correspond to a single quantum fluctuation. If, for example (see note 3), the electron correlates to two fluctuations and the proton to a single fluctuation, the estimated fluctuation mass density becomes (1836/3 + 2) x 9.93 x 10^{-29} grams /cc., or 0.61 x 10^{-29} grams/cc. This value is far more consistent with the observed "flat" universe than the previous QED estimate of 10^{122} grams/cc. With this value, vacuum fluctuations contribute about 2/3 of the universe energy density of about 0.9 x 10^{-29} grams/cc, as shown in **Figure 2**.

Conclusion:

Fundamental considerations provide a value for the vacuum energy density corresponding closely to that required for the observed "flat" universe. These considerations also show the accepted Planck Scale is seriously in error and the correct value corresponds to the naturally occurring particles. This analysis further indicates the source of an error of about 10^{42} within physics and illustrates how energy is localized to form particles.

The data presented in Figure 2 is in part derived from an article in the May 1988 issue of *Scientific American,* "The Mystery of the Cosmological Constant."

Note 1:

Singularity issues can be avoided by regarding the system as relativistic, precessing, and elliptical in nature, thereby preserving quantized angular momentum and an equivalent velocity c.

Figure 2: Allowed Values of the Cosmological Constant versus matter density, obtained by counting galaxies in regions of the universe and thereby determining the geometry of those regions. The gray area is a region of 95% confidence. Data from an analysis by Edwin D. Loh, based on work with Earl J. Spillar, both of Princeton University.

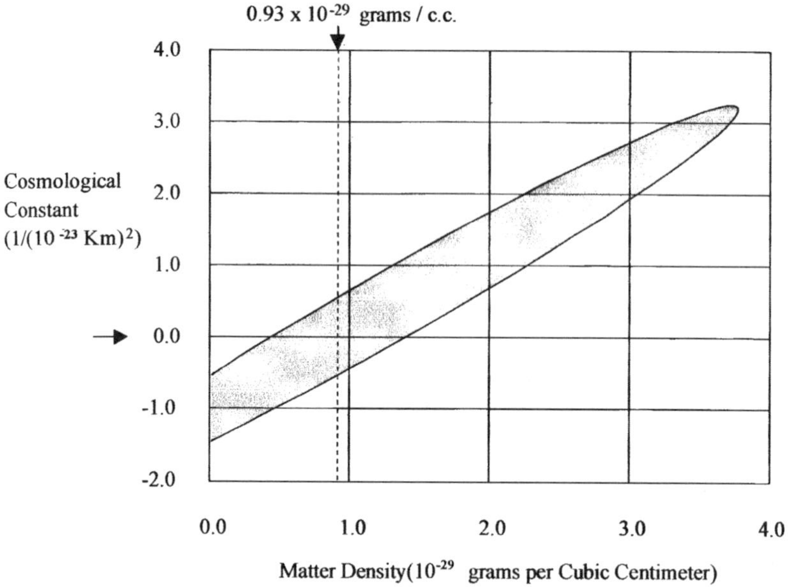

Note 2:

The nature of the "apparent force" F, constraining the mass in the "orbit" can be evaluated. From above, $F_S = \hbar c/r^2 = mc^2/r = G_S m^2/r^2$, where $G_S = \hbar c/m^2$. Rearranging $mc^2/r = G_S m^2/r^2$ gives $r = G_S m/c^2$. If $m = 2m_q$ say, then $r = 2G_S m_q/c^2$, the Schwarzchild radius (event horizon) of a "Black Hole" of mass m_q self-constrained under gravitational attraction with a force constant of G_S. All aspects of general relativity apply, hence the "apparent force" is simply a spherical metric curvature of radius r, just sufficient for the electromagnetic wave to propagate rectilinearly in the local metric, and hence Lorentz transformations apply. This suggests the strong force is no more than Super Gravitational attraction acting via mass, and elementary particles are sub-microscopic electromagnetic "Black Holes." Note the mass required for self-localization under Super Gravity is only m_q, half the electron mass, perhaps implying a two component electron.

Note 3:

By symmetry considerations, each mass distorts the metric spherically, hence gravitational interactions between two distant particles occur via two dimensions of metric curvature. The corrected "Planck mass" for gravitational interactions is 26.6 electron masses and relates to this two dimensional curvature, not the volumetric metric distortion. The full volumetric distortion clearly relates to $26.6^{3/2}$ electron masses or $137m_e$. The fine structure constant (α) is approximately 1/137. This numerical correlation is highly unlikely to be a coincidence, particularly as alpha is associated with the electron via the relation $e^2 = \alpha\hbar c$ (by Sommerfeld). The electron is therefore evidently a system in a 1/137 relativistic state with a mass in the rotating frame of about $137m_e$, i.e. 70 MeV/c^2. This is the value of the mass term in G_S. The internal mass (m_e/α) appears reduced by alpha (1/137) to the observed value of 0.511 MeV/c^2. Note that pions and muons are nominally at 140 and 105 MeV, suggesting increments of 35 MeV, further indicating the electron is a two-component system.

Note the proton mass at 938.3 MeV is greater than 35 MeV by about 26.6, suggesting a proton mass of approximately $m_q/\alpha^{2/3}$. The lack of integers or fractions in this value suggests the proton is a single component, although presumably distributed over three dimensions.

6.
More on the Proton, and Quarks

It is not the intent of this work to attempt development of a conceptual model for the proton, but only to point out that the ideas presented earlier are consistent with certain aspects of the current notion of quarks. It is noted that the proton mass at 938.3 MeV/c^2 is greater than the mass energy of 35.3 MeV/c^2 associated with the relativistic electron by about 26.58, suggesting a proton mass due to a relativistic state of approximately $m_q/\alpha^{2/3}$. The lack of integers or fractions in this value suggests the proton is a single component distributed over three dimensions. This at first appears to differ from the standard quark concept that represents the proton as a system of three quarks, in a u.u.d combination. However, a single constituent of 35.3 MeV distributed spherically necessarily has a component energy in each of the three orthogonal dimensions, and a three-component system is therefore not a contradiction as long as the components are properly defined.

Perhaps the defining characteristic of the quark concept is the notion of fractional charges, specifically 1/3 and 2/3 of the unit charge of classical physics. As the proton electric charge is +1, the opposite of the electron's charge of -1, the proton can perhaps be considered as a single component in a relativistic expanded state, compared to the electrons' compressed relativistic state. The electron is perhaps "collapsed" from an infinite radius, whereas the proton "expands" from a zero radius. Regardless of this inverse charge consideration, both the electron and proton are three-dimensional entities, presumably spherically distributed, so as in the quark concept we can assign a charge value to each dimension of the system as long as the total is plus or minus unity.

The two components of the electron are each of 35.3 MeV in an alpha relativistic state and consequently in the rotating frame of the mass have a "orbit" radius of 5.636 x 10^{-15} meters, twice the classical electron radius. In the observer's frame, the system is effectively rotating at c, so the measurable radius will be reduced to zero, and the electron will appear as a point particle. The proton, if an alpha$^{2/3}$ state, is in a significantly lower relativistic condition. The measured radius of the proton is about 1.09 x 10^{-15} meters, which is a factor of 5.15, or alpha $^{-1/3}$, less than that of the 35.3 MeV component as in the electron, which correlates to 5.62 x 10^{-15} meters or about twice the classical electron radius. It appears that both the proton and electron are comprised of half spin components of about 35 MeV in relativistic states relating to α, the fine structure constant.

Expressions describing various aspects of both the proton and electron apparently involve terms of $\alpha^{1/3}$ and $\alpha^{2/3}$. It does not take much imagination to correlate these fractional powers with the notion of 1/3 and 2/3 charges, particularly as $e^2 = \alpha\hbar c$. If a metric distortion constitutes a potential energy, it is clear that a volumetric distortion of $1/\alpha$ distributed equally over three dimensions is a distortion of $\alpha^{1/3}$ in each dimension. Hence if a particle of unity charge were considered on a dimension by dimension basis, then we would say there existed $\alpha^{1/3}$ of the charge in each dimension. In other words, the partial charges do not add linearly but as the product of partial powers of alpha, where the total charge value is given by the sum of the fractional powers.

Hence it appears that the above ideas produce the concept of 1/3 and 2/3 fractional charges via the dimensional distribution of the relativistic state, and thereby strongly support the quark-type notion as well as being consistent with the proton mass and the measured proton radius.

7.
The Final Chapter

So what of fundamental physics now? Lots of detailed work no doubt remains, but presuming the basis of this book is on target, far less now remains to be done at a philosophical level. If QED and Quantum GR (and string theory) can be quickly combined into a "Final Field Theory" with essentially no change to anything other than re-scaling GR, what remains to be done? Perhaps the answer is: beyond filling in the details, not much. The electron and proton can apparently be described as electromagnetic energy self-trapped in local space by metric curvature. But what about quarks and gluons and all the other exchange particles and stuff? Are they real or just imaginative figments of an era of ever increasingly complex mathematical "theory." If Super Gravity explains the "strong force," then do we need quantum-chromo-dynamics? Where do Z particles fit into the picture? The answer is: Sorry, I'm leaving all of those details for others to sort out.

The only pieces now missing from achieving a full understanding of the fundamental aspects of nature appear to be the basic properties of the metric that lead to the observed particles, along with some related questions. For example, why is the velocity of light (c) a particular value? Why does Planck's constant $\hbar$, have the value it has? Other than $\hbar$ and c, and the vacuum stress/strain and electric constants, are there any other fundamental parameters?

There are of course questions of a more cosmological nature, such as why is our universe full of matter instead of antimatter, what are the structures of the positron and anti-proton, and does any of this have to do with the age or size of the universe? Why do electrons and protons appear in apparently exactly equal numbers, etc.? Will it ever be possible to control gravity? Can antigravity exist? Is the Final Theory very close at hand? Will it take much longer to develop such a complete theory, or is the end of physics imminent? Perhaps completing the details on the issues addressed in this book will provide a window into these other issues; perhaps not, only time will tell.

Appendices

Appendix A

The meaning of $E = mc^2$ as representing a mass distortion m moving at velocity c in two dimensions simultaneously is illustrated below.

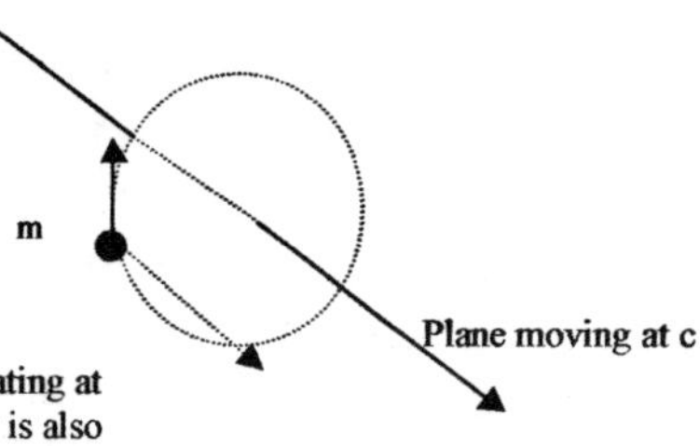

1) A mass distortion m rotating at velocity c in a plane that is also translating at c, is effectively moving at c in two dimensions simultaneously.

2) A mass distortion moving at velocity c in spherically curved space is effectively moving in two dimensions at c in each.

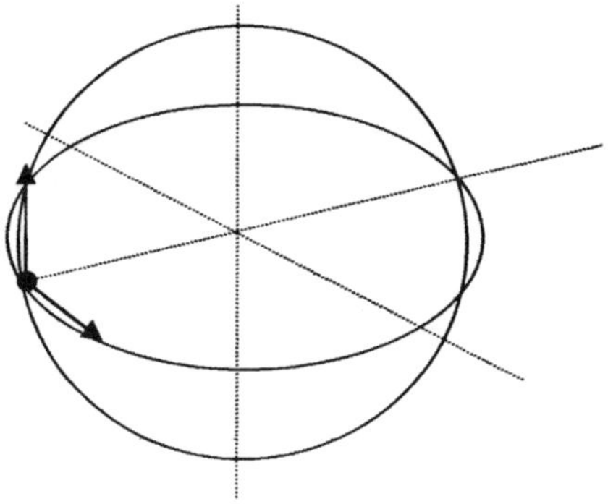

A mass M rotating at c about a radius r will logically be considered by the classical observer as constrained to orbit by a force $F = Mc^2/r$. If the mass energy is also given by $Mc^2 = \hbar c/r$, then the apparent force is $F = \hbar c/r^2$, and is at the strong force level. To the observer this represents a quantum black hole of energy Mc^2, apparently localized at the Strong Force level. The energy wavelength is equal to the orbit circumference so a quantum state exists, and a black hole exists as $r = \hbar c/Mc^2 = (\hbar c/M^2)M/c^2 = 2G_s(M/2)/c^2$, the radius of a black hole self trapped by Super Gravity acting with a force constant G_s.

The of Meaning of "Dimension"

The space-time domain in which the conscious mind of the observer exists is commonly presumed to be a 3D-1T frame, i.e., consisting of three spatial dimensions and one time dimension. In this context all dimensions are orthogonal i.e., a spatial dimension is defined as one that has a zero projection onto any other dimension.

By Einstein's theory of Special Relativity, a length L in a frame moving at velocity v to the observer appears to the observer as of length $L(1-v^2/c^2)^{1/2}$. Hence a length in a frame moving at a velocity $v = c$ has a zero length in the observer's frame and is therefore in another dimension. A mass moving at velocity c to the observer in a plane no longer has mass in the observer's domain but transforms to momentum $p = mc$. Similarly if the plane containing the moving mass is now translated at c, the momentum in the plane reduces to zero and the observer will perceive only energy, $E = pc = mc^2$.

Appendix B

Numerical value of Newton's classical gravitational constant based on the derived relation where the derived value of G_N is G_O, i.e., $\mathbf{G_O \equiv (1/\underline{c}^4)\hbar\, c/(m_e/\alpha^{2/3})^2}$.

With $\underline{c}$ always in cgs units, $\underline{c} = 2.9979 \times 10^{10}$, $\underline{c}^4 = 8.0773 \times 10^{41}$ and $[1/\underline{c}^4] = 1.2380 \times 10^{-42}$

Sample calculation in MKS units using following values:

m is the electron rest mass; $m = 9.1095 \times 10^{-31}$ kg, therefore $m^2 = 8.298\ 10^{-61}$ kg.

$\hbar = h/2\pi = 1.0546 \times 10^{-34}$ J/s and $c = 2.9979\ 10^8$ m/s, therefore $\hbar c = 3.1616 \times 10^{-26}$.

The fine Structure Constant, alpha = 1/137.04, hence alpha$^{2/3}$ = 0.03762 and alpha$^{4/3}$ = 1.4153×10^{-3}

$G_O = (1.2380 \times 10^{-42})(1.4153 \times 10^{-3})(3.1616 \times 10^{-26}) / (8.298 \times 10^{-61})$
$= (5.5391 \times 10^{-71}) / (8.298 \times 10^{-61})$
$= \mathbf{6.6758 \times 10^{-11}}$, in what heretofore was thought to be MKS units

Typical published value for empirically determined value of $\mathbf{G_N = 6.672 \times 10^{-11}}$, with a one sigma error of 4 in the last significant number.

Ratio $G_O/G_N = 1.00057$

Appendix C

Recent Developments in the Experimental Determination of G.

The value of G derived in this book is within the experimental range of uncertainty currently accepted for G. However, recent developments suggest the true uncertainty in the value of G is perhaps far greater than has generally been believed. The issue was brought to light by the April 1995 meeting of the American Physical Society in Washington D.C., as is described in an easily readable article listed below as Reference 5.

Historically, almost all measurements of G have been based on improved versions of the basic torsion balance technique used in 1798 by Henry Cavendish, an English nobleman. This consists of a 'dumbbell' of weights suspended horizontally by a thin filament attached at the dumbbell's center of gravity, with two heavy masses located close to the ends of the dumbbell. These masses attract the masses at the end of the dumbbell and thereby cause a small twist in the filamentary suspension. The magnitude of the induced twist permits a calculation of G via application of Newton's equation. By 1982 (see Ref. 1), data obtained using this and similar techniques were believed to have established the value of G to within an uncertainty of about 0.0128%.

However, more recent measurements by credible research teams in Russia, New Zealand, and Germany have obtained wildly differing values for G that now (May 2000) cast serious doubt on the value heretofore accepted. The German Institute of Standards has obtained a value 0.6% higher than the accepted value, and the University of Wuppertal in Germany has obtained a value 0.06% lower. The Russian results show various values differing by up to 0.7%, and the New Zealand Measurements Standards Laboratory found a value that is 0.1% lower. These disparate results have created a controversy over the value of G and have stimulated others to reconsider the issue. New efforts using somewhat different approaches are now underway at both the University of Washington and the University of California at Irvine. Perhaps the most promising new technique is discussed in the July 15, 2000 issue of Physical Review. Other references are provided below.

One historically disturbing aspect of experimental physics is that experimenters are logically compelled to seek endless "correcting factors" to their measurements until the theoretically predicted value is validated, at which time no more adjustments are needed and the search for additional "corrections" ends. In the case of G, however, there has been no theoretical value to aim for until now. The value of G given in Appendix B has not yet been publicized, and experimenters therefore feel uncertain about how to obtain closure or where the search will end.

1. G.G. Luther and W.R. Towler, Phys. Rev. Lett., 48, 121 (1982).
2. E.R. Cohen and B.N. Taylor, Rev. Mod. Phys., 59 1121 (1987).
3. W. Michaelis, H. Haars, and R. Augustin, Metrologica 32, 267 (1995).
4. M. Fitzgerald and T.R. Armstrong, IEEE Trans. On Inst. And meas. 44, 494 (1995).
5. H.C. von Baeyer, Discover Magazine, March 1996, P. 96-101.